Transforming your organization with G Suite
Tips and best practices

by Yoshiki Sato, Kim Wylie, Ayaka Yamada, Kenneth Siow, Shahid Nizami

Copyright ©Yoshiki Sato, Kim Wylie, Ayaka Yamada, Kenneth Siow, Shahid Nizami All rights reserved.
Printed in Japan
First Printing November 24th 2017, First edition (Version.1.0)
ISBN978-4-8443-9787-8

Editor-in-chief : Takashi Yamashiro
Digital Editor: Syo Kurihara
Publisher : Masanobu Iseri
Published by Impress R&D
Jinbocho Mitsui Bldg. 1-105 Kanda Jinbocho, Chiyoda-ku Tokyo 101-0051 Japan
https://nextpublishing.jp/

All rights reserved. No part of this publication may be reproduced, distributed, or transmitted in any form or by any means, including photocopying, recording, or other electronic or mechanical methods, without the prior written permission of the publisher, except in the case of brief quotations embodied in critical reviews and certain other noncommercial uses permitted by copyright law. For pe   ---ts write to the publisher, addressed above.

Prologue 6

Benefits of Transitioning to the Cloud 6
Reduce costs with cloud based systems 6
Exit the hardware maintenance cycle 7

What's the difference? Free vs. Subscription Cloud Services 9
Email with your domain 9
Centrally manage IT systems 10
Administrator support services 11

The Philosophy behind G Suite 12
100% Web Based Service 13
Simplified Networking 13
Minimal Downtime 13
Familiar Environment 14
Integration with other Google services 14
Transition mail from legacy systems 16
Offline support 18

An Overview of G Suite Services 19
Services that accelerate Communication 19
Services that accelerate Collaboration 23
Services that accelerate Administration 26
G Suite Edition 28

G Suite Customer Success Methodology 32
Digital transformation 32
G Suite Customer Success Methodology 33
Four Phases of Customer Success Methodology 34
Three phased deployment model 35

Chapter 1: Understanding G Suite Security 37

How Google protects G Suite security and privacy 37
Security is everyone's responsibility 37
Defense in depth at scale by default 38
Data center and infrastructure 39
Custom server hardware and software 39
End-to-end encryption by default 40
Endpoint Security in Every Layer 41
Third-party audits and certification 42
Data processing terms 43
Additional controls and capabilities 44

How G Suite empowers customers to improve

security and compliance ·· 45

User authentication/authorization features ······························· 46

Email Encryption ·· 47

Data Loss Prevention (DLP) ··· 48

Audit Logging ·· 49

eDiscovery and compliance ·· 50

Chapter 2: Getting Started with G Suite ···················52

Signing Up for G Suite·· 52

Custom Domain: Bring your own or purchase one!····················· 52

Sign up for a free 14-day trial ·· 53

Multiple methods to verify domain ownership ··························· 55

First Steps after Signing Up ·· 56

Creating new users·· 56

Verify domain ownership·· 57

Configuring your domain's MX records ···································· 60

Set up your payment information ·· 61

Techniques for Email Transition ······································ 62

Three Deployment Phases ··· 62

Core IT ··· 63

Early Adopters ·· 64

Global Go-Live··· 64

Data Migration from Legacy Systems ································ 65

Data migration led by system administrators ···························· 65

Data migration led by users ·· 68

Link your company's directory service to G Suite ······················ 69

Chapter 3: Change Management ··························71

Introducing the importance of Change Management···················· 71

What is Change Management? ··· 71

The importance of Change Management ··································· 72

Google's proven Change Management Methodology ·················· 73

Overview of the methodology·· 73

1. Sponsorship & Engagement: ··· 74

2. Organizational Analysis:··· 76

4

3. Communications: ⋯⋯⋯⋯⋯⋯⋯⋯⋯⋯⋯⋯⋯⋯⋯⋯⋯⋯⋯⋯ 77
4. Training: ⋯⋯⋯⋯⋯⋯⋯⋯⋯⋯⋯⋯⋯⋯⋯⋯⋯⋯⋯⋯⋯⋯⋯⋯⋯ 79
Additional activities ⋯⋯⋯⋯⋯⋯⋯⋯⋯⋯⋯⋯⋯⋯⋯⋯⋯⋯⋯⋯⋯ 80

Increasing adoption of G Suite post go-live ⋯⋯⋯⋯⋯⋯⋯ 80
The go-live date is just the beginning of the journey ⋯⋯⋯⋯⋯⋯ 80
Ongoing organisational analysis ⋯⋯⋯⋯⋯⋯⋯⋯⋯⋯⋯⋯⋯⋯⋯ 81
Facilitate Transformation Labs ⋯⋯⋯⋯⋯⋯⋯⋯⋯⋯⋯⋯⋯⋯⋯ 81
Establish an innovation council ⋯⋯⋯⋯⋯⋯⋯⋯⋯⋯⋯⋯⋯⋯⋯ 82
Continue with communications and training activities⋯⋯⋯⋯⋯⋯ 82
Identify where you are on your Transformation Journey ⋯⋯⋯⋯⋯ 83
Project execution ⋯⋯⋯⋯⋯⋯⋯⋯⋯⋯⋯⋯⋯⋯⋯⋯⋯⋯⋯⋯⋯ 84
Attributes of a good change manager ⋯⋯⋯⋯⋯⋯⋯⋯⋯⋯⋯⋯ 85

Chapter conclusion ⋯⋯⋯⋯⋯⋯⋯⋯⋯⋯⋯⋯⋯⋯⋯⋯⋯⋯⋯ 86
Closing note ⋯⋯⋯⋯⋯⋯⋯⋯⋯⋯⋯⋯⋯⋯⋯⋯⋯⋯⋯⋯⋯⋯⋯ 86

Chapter 4: Use cases from existing customers ⋯⋯⋯⋯⋯⋯⋯87

How companies are leveraging G suite to transform productivity ⋯⋯ 87
Security enhancement⋯⋯⋯⋯⋯⋯⋯⋯⋯⋯⋯⋯⋯⋯⋯⋯⋯⋯⋯⋯ 87
Legacy migration ⋯⋯⋯⋯⋯⋯⋯⋯⋯⋯⋯⋯⋯⋯⋯⋯⋯⋯⋯⋯⋯ 88
Global Communication Platform ⋯⋯⋯⋯⋯⋯⋯⋯⋯⋯⋯⋯⋯⋯⋯ 89
Sales Forecast with Google Sheets⋯⋯⋯⋯⋯⋯⋯⋯⋯⋯⋯⋯⋯⋯ 90
G Suite for Recruitment⋯⋯⋯⋯⋯⋯⋯⋯⋯⋯⋯⋯⋯⋯⋯⋯⋯⋯⋯ 91
Connecting with the organization socially ⋯⋯⋯⋯⋯⋯⋯⋯⋯⋯⋯ 91

Chapter 5: Merits of Deploying G Suite ⋯⋯⋯⋯⋯⋯⋯⋯⋯93

Business transformation with G Suite⋯⋯⋯⋯⋯⋯⋯⋯⋯⋯⋯ 93
Workstyle transformation with G Suite ⋯⋯⋯⋯⋯⋯⋯⋯⋯⋯⋯ 93
Efficient information sharing helps to achieve your business goals ⋯⋯⋯⋯ 94
Closing thoughts ⋯⋯⋯⋯⋯⋯⋯⋯⋯⋯⋯⋯⋯⋯⋯⋯⋯⋯⋯⋯⋯ 95

About the Author⋯⋯⋯⋯⋯⋯⋯⋯⋯⋯⋯⋯⋯⋯⋯⋯⋯⋯⋯ 98

Prologue

Benefits of Transitioning to the Cloud

||||||||||||||||||||||||||||||||||||||||||||||||||||||||||||||||||||||||||||||||||||||||||||||||||||||||||

Section Overview:

The "on-premise" model of keeping IT systems on site has multiple drawbacks. In comparison, what are the advantages of the cloud? How can your business benefit from transitioning to the cloud?

||||||||||||||||||||||||||||||||||||||||||||||||||||||||||||||||||||||||||||||||||||||||||||||||||||||||||

Reduce costs with cloud based systems

Traditionally, if you deployed an IT computer system (e.g. email or file sharing), you installed and maintained the system on the premises of your organization. This methodology (known as on-premise) has long been the industry standard.

Many businesses continue to embrace the on-premise model. Often, companies feel safer storing their data on hardware that is within an arm's reach, and therefore keep their systems on-premises. Many of these businesses also invest in systems designed to prevent unauthorized disclosure or loss of this data.

In comparison, organizations are increasingly shifting to cloud systems as an alternative deployment strategy with many benefits.

With the cloud, locally operated servers become unnecessary. You can transfer to the cloud at any time and transition as many or as few services as necessary. Furthermore, compared to the on-premise model, the time associated with installation and initialization is minimal, thereby reducing your costs.

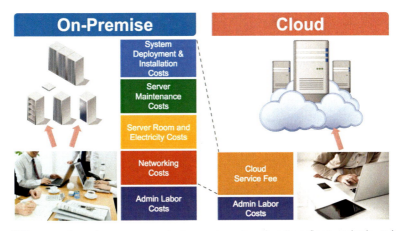

With on-premise systems, your organization must purchase, install, configure, maintain, and dispose of hardware. Because these steps are removed with the cloud, you can realize the cost savings.

Exit the hardware maintenance cycle

The hardware maintenance cycle of an on-premises system is a time-consuming and expensive process. Let's examine a typical on-premises hardware lifecycle to gain a better understanding.

First you must install the system on-site, a process which is usually complex and behind schedule due to system integrations and early failures. Next, you must conduct maintenance and upkeep. Because this is frequent and ongoing, you usually purchase a maintenance contract alongside the hardware, thereby increasing your total cost of ownership. Finally, upgrades near the end of the life cycle are costly and/or troublesome. Then, you repeat the cycle - indefinitely.

This is where cloud-based systems have a large advantage. With the cloud, you do not conduct any maintenance. Instead, the cloud service provider handles all operational upkeep and system maintenance.

Furthermore, you don't have to update your applications or security software as this is also done by the provider. By exiting the maintenance lifecycle, your business can utilize the newest technology, minimize security risks, and focus its attention on core functions.

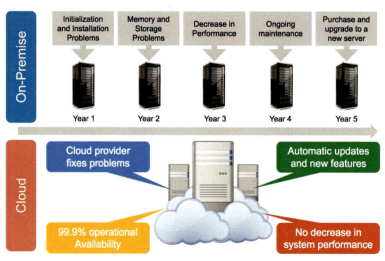

With on-premise systems, hardware and software are maintained by your system administrator. With cloud systems, maintenance is conducted by the cloud vendor.

What's the difference? Free vs. Subscription Cloud Services

||||||||||||||||||||||||||||||||||||||||||||||||||||||||||||||||||||||||||||||||||||||||||||||||||||

Section Overview

There are many cloud services freely available. Free cloud services have a lot to offer to businesses, so what's the difference between these and other subscription-based cloud services?

||||||||||||||||||||||||||||||||||||||||||||||||||||||||||||||||||||||||||||||||||||||||||||||||||||

Email with your domain

Google's flagship free service is its mail application Gmail. When you use Gmail for free, the standard address format is "example@gmail.com". Upon initiation, you are required to customize the "example" portion to a unique alias. However, you cannot change the domain portion of the address (@gmail.com). The domain "@gmail.com" is an indication that the system is hosted by the free version of Gmail.

Although "@gmail.com" may be fine for personal use, it may not be suitable for your organization. Many organizations own a domain, and therefore, want to use their existing domain as opposed to "@gmail.com" (For example, the publisher of this book uses the domain "@impress.co.jp").

If you want to use Gmail with your domain, you can do so by upgrading to G Suite (the subscription version of Google Cloud services, formerly Google Apps). With G Suite, during the initialization process, you can register an existing domain, or alternatively purchase and use a domain of your choice.

Prologue | 9

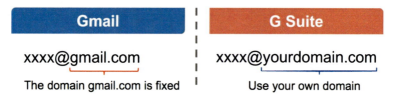

With the consumer version of Gmail, the domain must be "gmail.com" whereas with G Suite, a custom domain can be used.

Centrally manage IT systems

Within G Suite, administrators have access to a tool called the G Suite Admin Console. The Admin Console integrates all of G Suite admin services into a single service and support tool. With the Admin Console, you can easily manage an organization's users, services, and security settings.

Many IT security breaches involve employees unknowingly performing actions that expose themselves or their company. The Admin Console has many features available to prevent these types of employee-induced risks, as well as features designed to enforce a secure operating environment.

For example, using the Admin Console, you can enforce security measures such as password length, or restrict file sharing with external entities.

The Admin Console enables system administrators to manage settings easily from a central, web-based interface.

Administrator support services

  G Suite comes with a full suite of technical support services for your system administrators. Administrators have 24-hour access to support through email, phone, and online. Phone support is currently available in 14 languages, and online support in 18 languages, including English, Bahasa Indonesia, Thai, Mandarin, Korean, Spanish, German, and Japanese.

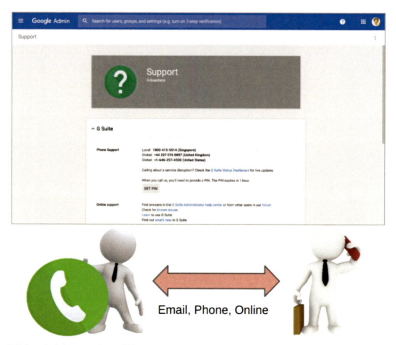

G Suite administrators have 24-hour access to support services through the Admin Console.

The Philosophy behind G Suite

IIIIIIIIIIIIIIIIIIIIIIIIIIIIIIIIIIIIIIIIIIIIIIIIIIIIIIIIIIIIIIIIIIIIIIIIIIIIIIIIIIIIII
Section Overview
G Suite is designed to let you work from "whenever", "wherever" , and from "whatever device". It was fundamentally designed this way to increase the productivity of Google's own employees. Let's take a look at how this can change the way your businesses works.
IIIIIIIIIIIIIIIIIIIIIIIIIIIIIIIIIIIIIIIIIIIIIIIIIIIIIIIIIIIIIIIIIIIIIIIIIIIIIIIIIIIIII

100% Web Based Service

With only a web browser, you can use all G Suite services. It is therefore, a 100% web-based cloud service. There is no need to install or maintain any software on your device beyond an internet browser. Furthermore, because G Suite is a 100% web-based, you can access your data from a PC, smartphone, and tablet.

Simplified Networking

G Suite architecture can help reduce the impact of the bandwidth usage of each services. For instance, when a user receive a message, Gmail downloads only the message headers, which are small relative to the total size of the message.

In addition, Gmail offers options to downloading file attachments that reduce bandwidth use. Users can see an in-line snapshot of an attached graphic and view a preview of an attached PDF and other common files in their browser. So you can access attachments without consuming as much bandwidth as downloading, and also avoid issues related to viruses embedded within these document types (e.g. macros that automatically run when opening a Word or Excel document) although the default protection from viruses built into Gmail and Google Drive.

Minimal Downtime

G Suite upholds 99.9 % operational and service availability through a Service Level Agreement (SLA) for all of its services. This means that there is virtually no downtime for your essential business activities. As evidence, Gmail's service level in 2013 was far beyond the minimum, at 99.978%.

Google can provide this SLA with no planned downtime for maintenance. Furthermore, Google has developed robust capabilities to mitigate against downtime due to natural disasters to ensure that the SLA is always upheld.

Familiar Environment

Gmail is one of the most popular email services, with a user base that has grown from 425 million in 2012 to over 1 billion in 2016 (announced at Google I/O 2012 and 2016 respectively).

As a result of this large user base, many individuals who use company-specific email and applications use Gmail at home. In those circumstances, training costs to implement Gmail can often be drastically reduced, as employees are already familiar with the functionality. The same applies to all G Suite services, not just Gmail.

Integration with other Google services

Many of Google's services are heavily integrated with G Suite. For example, messages you receive in Gmail can immediately be translated using Google Translate. With a single click in Google Calendar, you can investigate the location of a particular event in Google Maps.

An example of G Suite integrating with other Google services. With one click, Gmail messages can be translated using Google Translate.

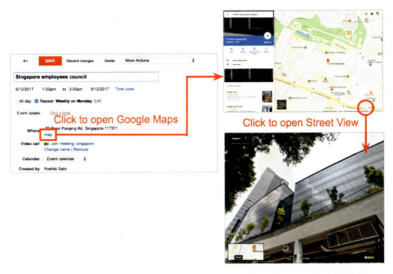

An example of G Suite integrating with other Google services. An event location in Google Calendar can be researched using Google Maps & Street View.

Furthermore, data contained in G Suites (including Gmail, Google Drive, and Google Hangouts) can be searched using Google's search capabilities. There is no need to search separately through Gmail, Drive files, or Hangout chats, as G Suites searches all data simultaneously. For

Prologue | 15

G Suite Business or Enterprise, Google Cloud Search is included as a core service of G Suite. This is a key capability of Google's cloud-based system and would be impossible with an "on-premise" model.

Transition mail from legacy systems

Many G Suites customers originally used a different mail service (e.g. Exchange Server or Lotus Notes). To migrate data from these services, Google has developed tools that you can use for free for G Suite users. For example, to transfer data from Exchange, you can use G Suite Migration for Microsoft Exchange. For Lotus Notes, you can use G Suite Migration for IBM Notes.

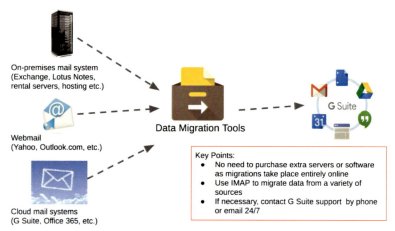

G Suite users can utilize various free migration tools to transfer their mail and associated data easily from legacy systems.

If your company uses Microsoft Outlook, your employees can transfer data themselves using GSMMO (G Suite Migration for Microsoft Outlook). If you want to use G Suite and Outlook simultaneously, you can sync your data using GSSMO (G Suite Sync for Microsoft Outlook).

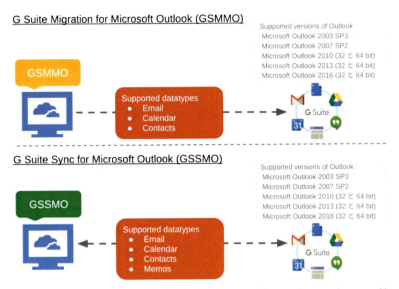

Free tools are available to transfer data from legacy systems. You can also sync data to enable simultaneous use.

In addition, if you use Microsoft Office (e.g. Word, Excel, PowerPoint), you can open and edit documents within G Suite using Google Drive and Google Documents. Google Documents is a word processing application that provides an experience and functionality similar to Microsoft products. For example, existing Microsoft Office documents can be transferred and edited using Google's word processing application known as Google Documents. Similarly, Excel formulas can be transferred and edited within Google Sheets.

Furthermore, Google offers a free add-on to the Chrome web browser which enables you to natively edit files formatted in Word,

Excel, and PowerPoint. Rather than perpetually purchasing expensive Microsoft Office licenses, transitioning light Office users (e.g. file viewers and minor editors) to G Suite is a logical choice and often leads to significant cost savings.

Using an add on to the Chrome browser, you can natively edit files formatted in Word, Excel, or PowerPoint.

Offline support

  G Suite provides an online support portal, which contains information such as common troubleshooting techniques or deployment best practices. If an application stops working, there are methods to report the error directly to Google, which helps Google's support teams identify and solve your problems.

	Mobile devices	Mobile devices	Laptops	Laptops
	Android and iOS	Android and iOS	Chrome browser	Chrome browser
	View	Edit	View	Edit
My Drive	✓	✓ (i)	✓	
Documents	✓	✓	✓	✓
Sheets	✓	✓	✓	✓
Slides	✓	✓	✓	✓
MS Office files	✓	✓ (ii)	✓	

(i) Editing limited to starring files and folders

(ii) Requires installation and use of QuickOffice

An Overview of G Suite Services

||||||||||||||||||||||||||||||||||||||||||||||||||||||||||||||||||||||||||||||||||||||||||||||

Section Overview

G Suite is packed with services that are designed to increase communication and collaboration. How can your business utilize these features to transform the workplace?

||||||||||||||||||||||||||||||||||||||||||||||||||||||||||||||||||||||||||||||||||||||||||||||

Services that accelerate Communication

Gmail is the flagship service of G Suite. The G Suite version of Gmail has a large single user storage limit of 30GB (shared with Google Drive). If you use G Suite Business or Enterprise, the storage is unlimited. Other features include Google's infamous search capability, which ensures quick and easy discovery, and its extensive spam blocking capabilities.

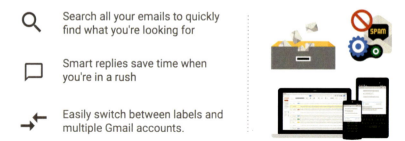

- Search all your emails to quickly find what you're looking for
- Smart replies save time when you're in a rush
- Easily switch between labels and multiple Gmail accounts.

In addition to email functions, Gmail is integrated with Google Hangouts, which enables you to check co-worker availability, text chat, and conduct video conferences.

Often used alongside Gmail, Google Calendar is another popular G Suite service. Google Calendar enables users to easily coordinate schedules, shares group calendars, book events, and even reserve common assets such as conference rooms or projectors.

- Schedule events quickly by checking coworkers' availability or layering their calendars in a single view
- Access from your laptop, tablet or Smart phone
- See if meeting rooms or shared resources are free

Google Calendar enables you to book events and easily share your schedule with others. Company assets such as conference rooms or projectors can be reserved directly from the calendar interface.

20 | Prologue

Another core communication tool within G Suite is Google Hangouts. Google Hangouts enables text-based chatting as well as voice and video conferencing for up to 25+ simultaneous users. Other essential features include the ability to check a user's communication availability in real time and the ability to screen share during presentations.

Create, edit and share docs from your iPhone, iPad or Android devices

Commenting, chat and real-time editing with your colleagues

Previous versions are kept indefinitely and they don't count toward storage

With Google Hangouts, you can text chat, conduct voice calls, video conferences, and screen share during presentations. A maximum of 25+ unique users can video conference in a single meeting.

Starting in 2014, "Chrome devices for Meetings" was released, equipped with the same Chrome Operating System (OS) that powers Google's Chromebooks. Each Chrome devices for Meetings unit is a plug and play video conferencing system, specifically designed to integrate with Google Hangouts. Priced at ~$1000 with a $250 annual management & support fee, it provides HD video conferencing for a significantly cheaper price than current systems.

Starting in 2017, "Jamboard" was released. It is Google's cloud-based, collaborative whiteboard. G Suite plan is required to use Jamboard so that you can access files from Drive, use them in your brainstorms and come back to your work later.

Chrome devices for meetings seamlessly integrates with Google Hangouts and provides HD video conferencing at a low price.

Jamboard is is the Most Powerful Whiteboard Ever. It's Google's collaboration tool and it's going to change how your teams grow ideas together.

Services that accelerate Collaboration

Google Drive is an online storage web application that has embedded word processing, spreadsheet editing, and presentation building services.

Because files saved in the Google Document format do not count towards the Google Drive storage quota, you are free to create, store, and save without restraint.

Furthermore, Google documents have a user interface similar to Microsoft Office, so you can quickly grasp how to view and edit documents.

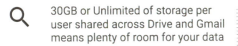
30GB or Unlimited of storage per user shared across Drive and Gmail means plenty of room for your data

Sync all or some of your files to a designated folder on all your devices

View over 40 popular file formats with Drive, including videos, images, Microsoft Office documents and PDFs

Google Drive is an online file storage and sync service. Along with powerful search and discovery functions, Google Drive enables users to access files from any smart device using only a web browser or the Drive App.

With Google Documents, a single document can be edited simultaneously by up to 50 different users. Edited documents can be saved in multiple file formats including Office, PDF, and txt.

🔍 Create, edit and share docs from your iPhone, iPad or Android devices

💬 Commenting, chat and real-time editing with your colleagues

↔ Previous versions are kept indefinitely and they don't count toward storage

Google Documents can be accessed by a maximum of 200 users and edited simultaneously by 50 users. Furthermore, Google Documents saved to Drive do not count against Google Drive's storage limit.

Files and documents stored on Google Drive can be shared with other users both internal and external to an organization. If necessary, administrators can adjust Drive file share settings using the Admin Console.

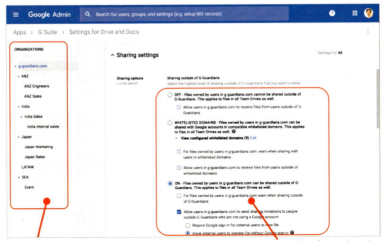

Apply unique settings for each organization

Allow / Restrict sharing outside of your domain

The Admin Console enables you to set security policies and ensure a safe operating environment. Policies can be applied at the domain level or only to a subset of organizations.

Another feature frequently used by businesses is Google Sites. Sites enables you to easily launch an intranet or a portal website that can be accessed by your employees. Common use cases include bulletin boards to share news and updates, or file cabinets to store essential documents.

Sites are built using a drag and drop interface. Therefore, no prior web development knowledge is necessary. Information from Google Documents and YouTube can be embedded directly into the Google Site.

- Create sites as simply as writing a doc through an intuitive editor
- Develop team sites with important content from projects, sales materials or details about the company picnic
- Enhance your site by embedding calendars, maps, videos, sheets, presentations and more

With Google Sites, you can build websites and intranet services such as company bulletin boards, portal sites, or project sites. No programming experience is necessary to build a Google Site.

Services that accelerate Administration

The Admin Console enables administrators to centrally manage user and application settings. From the Dashboard, you can navigate to different menus such as application settings, security settings, or support services.

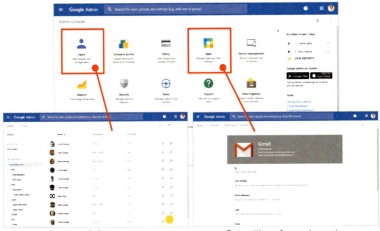

Manage G Suite users   Set settings for each service

The G Suite Admin Console is a centralized management tool that enables admins to efficiently and securely manage users and service settings.

The "Device Management" console enables you to centrally manage devices connected to your G Suite account, including Android, iPhone, iPad, and Windows Phone.

Beyond simple approval / denial settings for gaining access to G Suite, the Device Management Console can enforce security policies such as mandatory password protection and device encryption.

Furthermore, if a device is lost or stolen, you to remotely lock or wipe the device. The remote wipe function can either remove all information stored on the device or just information tied to the G Suite account.

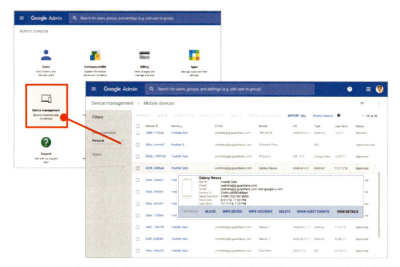

Within the Admin Console, the Device Management Console enables you to control which devices can access your corporate data. Administrators can enforce a variety of device level restrictions (e.g., no camera or external storage) as well as security settings such as password protection/encryption.

G Suite Edition

When you sign up for G Suite, you can choose among three options.

G Suite Basic gives you a professional email address at your company, 30 GB storage per person, and Google's suite of collaborative business apps—including Gmail, Calendar, Drive, Docs editors, Hangouts, Google+, Keep and more.

G Suite Business provides an enhanced office suite with additional features built specifically for the needs of businesses, such as advanced mobile device management and Google Vault.

G Suite Enterprise delivers a premium office suite that adds advanced controls and features for the most demanding enterprises. Features such as data loss prevention (DLP) and email Secure/ Multipurpose Internet Mail Extensions (S/MIME) encryption.

Basic services and features

	Basic	Business	Enterprise
Messaging: Gmail, Calendar, Contacts	✓	✓	✓
Storage and collaboration: Drive, Docs, Hangouts	✓	✓	✓
Web forums and shared inboxes: Groups for Business	✓	✓	✓
Domain-wide mail and document search, email retention: Vault	*	✓	✓
Other Google services: Blogger, YouTube, and more	✓	✓	✓
Unified search and assist for content in various G Suite services: Google Cloud Search		✓	✓

*   Google Vault is available with G Suite Basic as a paid add-on

Usage and support

	Basic	Business	Enterprise
Maximum number of users	Unlimited	Unlimited	Unlimited
Email, document, photo storage per user	30GB	Unlimited *	Unlimited *
99.9% uptime guarantee	✓	✓	✓
24/7 phone support	✓	✓	✓
Priority email support	✓	✓	✓
Video conferencing	✓	✓	✓

*   Accounts with 4 or fewer users get 1 TB per user

Security and business controls

	Basic	Business	Enterprise
Password monitoring and strength control	✓	✓	✓
2-step verification	✓	✓	✓
Single Sign On (SSO)	✓	✓	✓
Enforced SSL connections	✓	✓	✓
User-managed security keys	✓	✓	✓
Domain-wide admin managed security keys		✓	✓
Mobile device management	✓	✓	✓
Mobile device audit		✓	✓
Advanced Drive auditing and reports		✓	✓
Advanced Drive administrator controls		✓	✓
Drive data loss prevention			✓
Gmail data loss prevention			✓
Gmail log search in BigQuery			✓
Gmail integration with third-party archiving tools			✓
Deep Scanning Gmail Attachment (Zero day attack, Ransomware Protection, etc)			✓
Security key management			✓
S/MIME encryption for email			✓
Maximum number auto provisioning applications	3	Unlimited	Unlimited

Your Company brand

	Basic	Business	Enterprise
Your custom email address (you@yourcompany.com)	✓	✓	✓
Your www address in Google Sites (www.yourcompany.com)	✓	✓	✓
Ad-free experience	✓	✓	✓
Addresses at multiple domains (you@company2.com)	✓	✓	✓

Contacts, lists, and groups

	Basic	Business	Enterprise
Mailing lists (admin controlled)	✓	✓	✓
Mailing lists (user controlled)	✓	✓	✓
Global directory (internal directory)	✓	✓	✓
Global directory (external contacts)	✓	✓	✓
Forums and collaborative inboxes	✓	✓	✓

Access options

	Basic	Business	Enterprise
Supported browsers	✓	✓	✓
Mobile phones and tablets	✓	✓	✓
Offline mail and Docs editors	✓	✓	✓
IMAP and POP mail support	✓	✓	✓
Microsoft Outlook support	✓	✓	✓

Migration services

	Basic	Business	Enterprise
Migration tools and services	✓	✓	✓
Mail, contacts and calendar migration from Microsoft Exchange, Office 365, IBM Notes, and other webmail hosts	✓	✓	✓
Email import from webmail host	✓	✓	✓
Multi-user import from Exchange	✓	✓	✓
Multi-user import from IBM Notes	✓	✓	✓

Enterprise sync options

	Basic	Business	Enterprise
Sync user data with your LDAP server	✓	✓	✓
Use Exchange Calendar with G Suite	✓	✓	✓

G Suite Customer Success Methodology

||||||||||||||||||||||||||||||||||||||||||||||||||||||||||||||||||||||||||||||||||||||||||||||||||

Section Overview

Deploying G Suite is not just an IT project, but rather a project that needs to be aligned with customer's overall business goals. This section introduces the overview of Google's G Suite Customer Success Methodology.

||||||||||||||||||||||||||||||||||||||||||||||||||||||||||||||||||||||||||||||||||||||||||||||||||

Digital transformation

We believe that true "innovation" is only achieved when all three components, "People - Structured approach to support users for successful transformation", "Process - Keep on reinventing existing business processes with G Suite" and "Technology - Implement G Suite following our best practices", are acted together in a sustainable way. If an organization invests time and resources on these three components, G Suite adoption level increases gradually and as a result, an organization achieves its business and cultural transformation goals (See below for the adoption curve).

Transformation requires a prescribed and deliberate investment over time.

G Suite Customer Success Methodology

Based on our experience of deploying G Suite to over 3 millions organizations globally, we have developed the G Suite Customer Success Methodology. There are four stages that a customer goes through to fully gain the benefits of implementing G Suite;

Engage stakeholders	Deploy services	Grow adoption	Measure and promote successes
Build sponsorship, define scope and plan your support model, communications and training	Deploy services and execute on support model, training and communications plans	Establish a Google team, run Transformation Labs, and manage support operations	Measure adoption progress, promote success stories and support users through self help

Four Phases of Customer Success Methodology

There are five work streams (See below) that are essential to ensure customer success and help customers reach their 'Embed' lifecycle. Our recommendation is to assign a lead for each workstream (ie. Project Management lead, Technical lead, Change Management lead) and make sure the smooth progress of each workstream as well as alignment with other workstreams. We talk about People and Process work streams in the Chapter 3. Also, we touch on the most essential elements of Technology and Support work streams in Chapter 2.

	EXCITE	ENABLE	EXPAND	EMBED
Governance	Plan deployment	Manage implementation	Build a Google Team	Manage on-going program
People	Engage stakeholders	Execute training & communications	Conduct ongoing training & communications	Promote successes
Process	Determine business objectives	Initiate business process discovery	Implement business process solutions	Document and measure process improvements
Technology	Set technical scope	Deploy services	Stabilize platform	Grow Google platform
Support		Launch support model	Optimize end user support	Track trends & metrics
	SALES	DEPLOYMENT ~ 90 DAYS	ONGOING SUCCESS	

Five work streams of Customer Success Methodology

Three phased deployment model

Deployment is defined as the time that the customer signs the contract to the time the company goes live with G Suite. The standard time for deployment is 3 months, but we typically see this span longer for larger, more complex customers.

Deployment is broken down into three stages. Core IT: Prepare and train the IT team that is managing the journey. Early Adopters: Train and prepare employees who volunteer to provide peer support, who we call Change Champions. Global Go Live: Entire organization goes Google. We will use these phase names throughout this book.

Three phases of G Suite deployment

Chapter 1: Understanding G Suite Security

How Google protects G Suite security and privacy

||||||||||||||||||||||||||||||||||||||||||||||||||||||||||||||||||||||||||||||||||||||||||||

Section Overview

Google has spent the past 15+ years designing, building, and operating a highly secure infrastructure. Google has 7 consumer services with more than one billion daily users that run on that infrastructure, and Google Cloud allows other organizations to do the same.

||||||||||||||||||||||||||||||||||||||||||||||||||||||||||||||||||||||||||||||||||||||||||||

Security is everyone's responsibility

No single person is responsible for the security. It is the responsibility of the whole to ensure the privacy and accuracy of the information. How can Google Cloud help organizations with their security?

First, Google gives them a secure foundation. Google's core infrastructure is designed, built, and operated with security in mind.

Second, Google provides control and visibility. They are constantly expanding array of controls customers can use to help meet policy, regulatory, and business objectives.

Third, Google continually innovate applying expertise and investments to raise the security bar for Google, customers, and the industry.

Defense in depth at scale by default

Google's infrastructure doesn't rely on any single technology to make it secure. Rather, Google builds security through progressive layers that deliver true defense in depth.

The hardware is Google controlled, built and hardened. Any application binary that runs on our infrastructure is deployed securely. Google does not assume any trust between services, and use multiple mechanisms to establish and maintain trust - Google's infrastructure was designed to be multi-tenant from the start. All identities, users and services, are strongly authenticated. Data stored on our infrastructure is automatically encrypted at rest and distributed for availability and reliability. Communications over the Internet to our cloud services are encrypted. The scale of our infrastructure allows us to absorb many Denial of Service (DoS) attacks, and Google has multiple layers protection that further reduce the risk of any DoS impact. Finally, Google's operations teams detect threats and respond to incidents 24 x 7 x 365.

Google builds security through progressive layers that deliver true defense in depth.

38 | Chapter 1: Understanding G Suite Security

Data center and infrastructure

Google operates one of the largest backbone networks in the world, connecting data centers with hundreds of thousands of miles of fiber optic cable. Third parties estimate that more than 25% of global internet traffic flows over Google's network in a given day. Google has more than 100 points of presence across 33 countries and Google continues adding and scaling zones and regions to meet customers' preferences and policy requirements.

Google's network delivers low latency but also improves security. Once customers' traffic is on Google's network, it is no longer transiting the public internet, making it less likely to be attacked, intercepted, or manipulated.

Custom server hardware and software

Google's hardware infrastructure is custom-designed by Google "from chip to chiller" to precisely meet our requirements, including security.

Google's servers and the OS are designed for the sole purpose of providing Google services. Google's servers are custom built and don't include unnecessary components like video cards or peripheral interconnects that can introduce vulnerabilities.

The same goes for software, including low-level software and our OS, which is a stripped-down, hardened version of Linux. Further, we design and include hardware specifically for security - like Titan - Google's custom security chip that Google uses to establish a hardware root of trust in our servers and peripherals.

Google builds their own network hardware and software to

improve performance as well as security. This all rolls up to our custom data center designs, which include multiple layers of physical and logical protection.

Understanding provenance from the bottom of our hardware stack to the top allows us to control the underpinnings of our security posture. Unlike other cloud providers, Google has greatly reduced the "vendor in the middle problem" - if a vulnerability is found, Google can take steps immediately to develop and roll out a fix. This level of control results in greatly reduced exposure for us and our customers.

Purpose-built chips Purpose-built servers Purpose-built storage Purpose-built network Purpose-built data centers

Google data center consists of thousands of server machines connected to a local network. Both the server boards and the networking equipment are custom-designed by Google.

End-to-end encryption by default

Communications over the Internet to Google Cloud require properly terminated TLS connections. Once data reaches our infrastructure, it's encrypted at rest - by default with no extra effort.

Data is first "chunked" - broken up into pieces, and each chunk is encrypted with its own data encryption key. Each data encryption key is wrapped using a key encryption key for another layer of protection. The encrypted chunks and wrapped encryption keys are then distributed across Google's storage infrastructure. When data needs to be retrieved, the process repeats in reverse.

As a result, if an attacker were to compromise an individual key or gain physical access to storage, they would still be unable to read customer data.

User's data that is uploaded or created in G Suite services is encrypted at rest.

Endpoint Security in Every Layer

Google provides endpoints that follow a multi-layered security approach - enterprise devices based on Android and Chrome OS, as well as enterprise versions of the Chrome browser for cloud application access.

Devices benefit from universally available, tightly integrated security services to protect users from potentially harmful applications, strong on-device security primitives like application sandboxing and secure boot, and security programs like our bug bounty and vulnerability rewards programs.

Chrome browser provides malware and phishing protection,

automatic updates, and security settings and enterprise controls that sync across devices to ensure security policies are applied uniformly.

Google provides endpoints security in every layer.

Third-party audits and certification

Google's customers and regulators expect independent verification of our security, privacy and compliance controls. Google undergoes independent third party audits on a regular basis to provide this assurance. This means that an independent auditor has examined the controls present in our data centers, infrastructure and operations.

Google's certifications include the most widely recognized, internationally accepted independent security standards, including ISO 27001 for security controls, ISO 27017 for cloud security, and ISO 27018 for cloud privacy, as well as AICPA SOC 1, 2, and 3. These certifications help us meet the demands of industry standards such as CSA STAR and PCI-DSS, and many other regional standards as well.

Google infrastructure is certified for a growing number of compliance standards and controls.

Data processing terms

Google believes that trust is created through transparency, so Google's data processing and security terms for Google Cloud are available for anyone to review on the website.

Google offers customers a detailed data processing amendment that describes our commitment to protecting their data. It states that Google will not process data for any purpose other than to fulfill our contractual obligations. Furthermore, if customers delete their data, Google commits to deleting it from our systems within 180 days.

Google also describe the security measures we take and implement. Google commits to regular third-party audits as we described previously, as well as prompt notification of security incidents that we become aware of. Google discloses the use of any sub-processors.

Finally, Google provides tools that make it easy for customers to take their data with them if they choose to stop using Google's services, without penalty or additional cost imposed by Google.

Everyone can review Google's data processing and security terms.

Additional controls and capabilities

The security products which Google offers across Google Cloud help customers meet policy, regulatory, and business objectives. They include Identity and Access Management, Network Security, Data Protection and Governance, Application Security, Endpoint Security and Management, and Logging and Auditing. This rich set of controls and capabilities continues to expand over time.

		G Suite		
Identity & Access Management	✔	✔		
Network Security	✔			
Data Protection and Governance	✔	✔	✔	
Application Security	✔	✔	✔	
Endpoint Security and Mgmt.		✔	✔	
Logging and Auditing	✔	✔	✔	

Google helps customers meet their policy, regulatory, and business objectives.

How G Suite empowers customers to improve security and compliance

IIIIIIIIIIIIIIIIIIIIIIIIIIIIIIIIIIIIIIIIIIIIIIIIIIIIIIIIIIIIIIIIIIIIIIIIIIIIIII
Section Overview
Google builds security into its structure, technology, operations and approach to customer data. The robust security infrastructure and systems become the default for each and every G Suite customer. Beyond these levels, users are actively empowered to enhance and customize their individual security settings to meet their business needs.
IIIIIIIIIIIIIIIIIIIIIIIIIIIIIIIIIIIIIIIIIIIIIIIIIIIIIIIIIIIIIIIIIIIIIIIIIIIIIII

User authentication/authorization features

As Google has expanded beyond G Suite, the enterprise story gets richer, identity is front and center for all Google's services. Google is providing an Identity as a Service offering, called Cloud Identity. It made available to Google's customers even if they are not a G Suite customer.

Cloud Identity

Cloud Identity is an Identity as a Service (IDaaS) solution. It offers the Identity services that are available in G Suite as a stand-alone product.

Security Key was specifically designed to address the issues with One time password (OTP)-based 2-Step Verification and has been standardized in the FIDO Alliance under the technical standard "U2F", or Universal Second Factor.

These devices are designed in such a way that they are inherently safe against phishing attacks even if an attacker tricks the user into exercising this device on a phishing website, the resultant signature will not pass validation on the server, and access would not be granted to the attacker.

This is unlike current OTP-based authentication solutions which are still susceptible to phishing.

For G Suite Enterprise customers, Google added Security Key enforcement support from G Suite admin console. G Suite administrator can allow, disallow and monitor the Security Key usage for G Suite users. Also, they can enforce and revoke Security Key if necessary.

Methods for 2-step verification:

Verification method	Software or hardware	Requirements
Text message	Software	Cellular service and a powered mobile device
Google Authenticator	Software	Powered mobile device
Security keys	Hardware	Google Chrome desktop browser (version 40+), iOS, Android

Email Encryption

Google provides a hosted S/MIME service extending encryption capabilities on Gmail beyond TLS. TLS only guarantees to the sender's service that the first hop transmission is encrypted and to the recipient that the last hop was encrypted. But in practice, emails often take many hops (through forwarders, mailing lists, relays, appliances, etc). With hosted S/MIME, the message itself is encrypted. This facilitates secure transit all the way down to the recipient's mailbox.

S/MIME adds account-level signature authentication, which is unlike DKIM, which provides only domain-based authentication. This means that email receivers can ensure that incoming email is actually from the sending account, not just a matching domain, and that the message has not been tampered with after it was sent.

Admin or End-User uploads
S/MIME certs to Gmail Servers

End-User exchanges public keys with
recipient in their first email exchange

Future email is signed and
encrypted during transport

G Suite admin can enhance the integrity and confidentiality of their organization's email messages by enabling Secure/Multipurpose Internet Mail Extensions (S/MIME).

Data Loss Prevention (DLP)

G Suite's DLP protection goes beyond standard DLP with easy-to-configure rules and OCR recognition of content stored in images so admins can easily enforce policies and control how data is shared.

DLP automatically checks all email in Gmail, contents in Google Drive and mobile device events according to policies set by the G Suite admin and prompts the appropriate action.

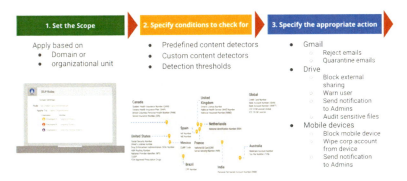

DLP for G Suite checks all email in Gmail, contents in Google Drive and mobile device events.

Audit Logging

G Suite Business and Enterprise provide a robust audit trail for G Suite customers. For G Suite Basic, audit logs are available for Admin activities, Logins, SAML, Calendar activities, Tokens, Groups, Google+ and Email log search. For G Suite Business and Enterprise, G Suite admins are also able to access audit logs for Google Drive and Mobile devices.

G Suite admins do have access to some of the Audit Logs via the Reports API. Having API access is an important feature as it allows companies to move this into unified logging systems to create a holistic view of user activity.

Audit logs contain valuable information that can help administrators diagnose issues or unlock insights. G Suite admin has a capability to analyze Gmail logs with a preconfigured BigQuery integration so they can run sophisticated, high-performing custom queries, analyze their data and build custom dashboards.

G Suite admin can find their business and IT insight from audit log data.

eDiscovery and compliance

Google Vault helps G Suite customers' organization meet its legal needs, by allowing them to manage their employees' G Suite data for eDiscovery and compliance purposes for Gmail, Drive, Team Drives and Google Groups.

To set retention policies, G Suite admin can place legal holds on their employees' Gmail, Google Drive and Team Drives files. If a user on hold deletes data, it will appear as deleted for him or her, but it is still available in Vault until the hold is removed.

Google Vault also allows G Suite admin to export revisions of their employees' Gmail, Drive and Team Drive files from a specific point in time. This can be done by simply specifying the desired Version Date in the search form.

Google Vault works with Groups, meaning G Suite admin can search, export, and set retention policies and place legal holds on their employees' Groups content. Groups can be used for email lists, forums, and shared or collaborative inboxes, and G Suite admin can apply the same retention and eDiscovery programs that they use in Gmail for content stored in Groups archives.

Google Vault is an e-discovery and archiving tool included with G Suite. G Suite admin can use Google Vault to comply with state and federal data retention regulations, monitor users' activity, and retrieve lost or destroyed user data.

Chapter 2: Getting Started with G Suite

Signing Up for G Suite

||||||||||||||||||||||||||||||||||||||||||||||||||||||||||||||||||||||||||||||||||||||||||||||||||||||||
Section Overview
G Suite is a 100% cloud-based-service; there's no need to install any software or servers. So what does your organization need to get started with G Suite?
||||||||||||||||||||||||||||||||||||||||||||||||||||||||||||||||||||||||||||||||||||||||||||||||||||||||

Custom Domain: Bring your own or purchase one!

G Suite enables you to use Google services with a custom domain. There are two types of custom domains:

1) Existing domains: owned prior to the initiation of G Suite

2) New domains: purchased during the initiation of G Suite

If your company already has a website or utilizes domain email services, you would use option 1 to register your existing domain with G Suite.

On the other hand, if you are a startup or small company that is just beginning to develop your online presence, you may use option 2 to purchase a new domain to use with G Suite.

Google makes it easy to try G Suite with a free trial. Existing domains, as well as new domains, can be used during the trial period.

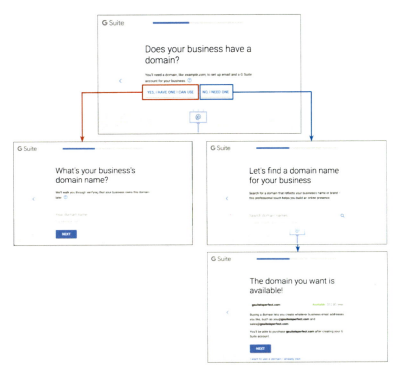

The G Suite free trial enables companies to use their existing domains or purchase a new domain during the initiation process.

The type of custom domain (existing or new), has no effect on the functionality of G Suite. If you decide to purchase a new domain from Google, step by step instructions are provided to navigate the process as shown in the figure above.

Sign up for a free 14-day trial

To get started with G Suite, you need to own a domain and follow the process outlined below:

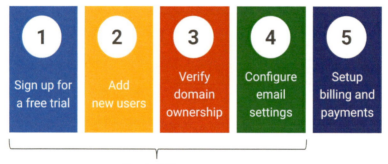

Required Steps

To get started with G Suite, you will need to register for a G Suite 14-day free trial and set up various parts of their G Suite account.

You can navigate to the G Suite homepage and follow the instructions below to sign up for a free trial. The free trials can be accessed from the G Suite homepages:

Address for G Suite homepage:
https://gsuite.google.com/

From the G Suite homepage, click "Get Started" to start the free trial enrollment process. The enrollment process consists of three steps:

1. Fill out your business and contact information
2. Choose whether to use your own domain or purchase a new domain
3. Create a G Suite account (this will be your admin account). Fill out the required information and Sign Up to finish

Multiple methods to verify domain ownership

If you are registering an existing domain with G Suite, Google will need to verify that you own the domain. This portion of the sign up process can be a bit technically involved but easily accomplished using the directions below.

There are four different methods to verify domain ownership. The following paragraphs will outline the general flow for the verification process. For detailed explanations and screenshots of each verification method, please consult the relevant figure on the following pages.

You first choose a verification method during the sign up process. A domain verification wizard will then help you input the necessary information and complete the verification. If at any point in the process you need help, you can receive guidance over the phone from Google's technical support services.

You only need to complete one of the following methods to verify a domain:

	Methods to Verify Domain Ownership	Verification Time
Method 1	Add a TXT record to your domain's DNS settings	Up to 1 hr
Method 2	Add a CNAME record to your domain's DNS settings	Up to 1 hr
Method 3	Upload an HTML file to your domain's web server	Typically right away
Method 4	Add a meta tag to your domain home page	Typically right away

First Steps after Signing Up

||||||||||||||||||||||||||||||||||||||||||||||||||||||||||||||||||||||||||||||||||||||||||

Section Overview

Because G Suite is a 100% cloud-based service, the initialization and setup are also done entirely online. This section walks through the initialization process using a step-by-step approach.

||||||||||||||||||||||||||||||||||||||||||||||||||||||||||||||||||||||||||||||||||||||||||

Creating new users

During the sign up process for the free trial, your system admin will set up their own user profile. However, to enable G Suite access for other members of the organization, the admin will need to add new user accounts.

Users can be added at any time to fit an organization's needs. There are three methods for adding users:

1. Individually add users through the Admin Console
2. Add users in bulk using a CSV file (up to 500 users per file)
3. Sync users from an Active Directory or LDAP server

Most readers of this book are likely to utilize method 1 as their primary method for managing users. However, if you need to add users in bulk, method 2 will likely become your primary choice. Finally, method 3 is necessary if your organization is transferring data from a legacy system.

Verify domain ownership

After setting up a free trial, the first thing you need to do is verify domain ownership. This process is necessary to ensure that you (the individual who registered for the free trial) are the owner of the domain. In other words, this prevents your business's domain from being used by another entity.

With G Suite, there are four ways to verify domain ownership. This example will walk through method 1, which is to add a TXT record. Let's see what steps need to be taken, as well as what Google is doing to verify ownership.

First, let's define what a TXT record is. A TXT is a type of DNS record which provides various text-based information associated with a domain. Interestingly, the IETF (an organization that promotes

Internet standards) does not define a specific use for TXT records and, therefore, TXT records are used to store information for a number of different purposes.

G Suite verifies domain ownership by providing you with a security token and checking to see whether the security token has been added to a TXT record on your domain's DNS records. In other words, Google believes you have domain ownership if you can add the specific TXT record.

Below is the domain ownership verification process for adding a TXT record:

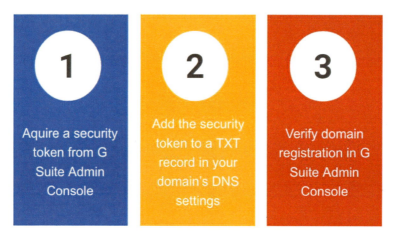

G Suite Admins acquire a security token from the G Suite Admin Console, then add the security token to a TXT record on the domain's DNS server. G Suite Admins can check to see if Google has verified domain ownership using the G Suite Admin Console.

After the verification process is complete, you will be able to send emails using their registered domain. However, you will still be unable to receive emails at this domain! Let's investigate why and determine which settings need to be adjusted.

At this point, you will only be able to receive emails from a "test domain" that Google has configured. This is because your MX records are continuing to route mail to your legacy system (more on this below). The Google test domain can be found in the G Suite Admin Console under the "Domain" menu and has the following naming convention: "yourdomain.test-google-a.com".

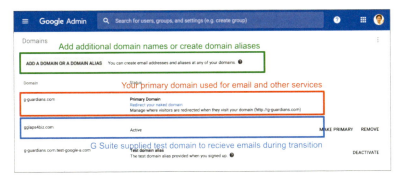

Two domains are shown under the domain menu in the G Suite Admin Console: "yourdomain.com" and "yourdomain.test-google-a.com". At this point, you can only receive mail at the test domain address.

If you want to check that your domain is properly registered with G Suite, you can send an email to the test domain address. In addition, by configuring your current mail server to forward emails to your test domain, you can receive emails at your current system as well as Gmail.

However, to directly receive emails sent to "yourdomain.com" in Gmail, you will need to configure your domain's MX records to Google's mail servers (see section below).

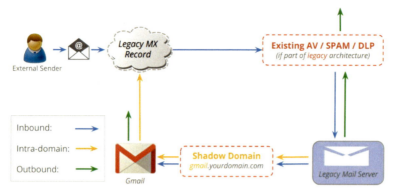

By forwarding emails sent to "yourdomain.com" to "gmail.yourdomain.com", you can receive emails in both your legacy server as well as Gmail.

Configuring your domain's MX records

MX records are special DNS records that direct a domain's mail flow. Each MX record directs mail to a specific mail server. By changing your domain's MX records, you can route your mail directly to Google's mail server, instead of your legacy server. The general process is very similar to the TXT record modification that we discussed earlier.

Changes to MX records can take up to up 72 hours to go into effect, although this usually occurs much sooner. Once the change is in place, mail sent to your registered domain will be received in Gmail.

A word of caution: Once your domain's MX record changes go into effect, mail will no longer be directed to your legacy mail server.

G Suite MX record values

Priority	Value/Answer/Destination
1	ASPMX.L.GOOGLE.COM
5	ALT1.ASPMX.L.GOOGLE.COM
5	ALT2.ASPMX.L.GOOGLE.COM
10	ALT3.ASPMX.L.GOOGLE.COM
10	ALT4.ASPMX.L.GOOGLE.COM

Set up your payment information

The time limit for a free trial is 14 days (with no option to extend). To continue using G Suite beyond the free trial, you will need to set up your primary payment method. There are two primary ways to purchase G Suite:

1. Purchase directly from Google using the G Suite Admin Console. Payments are done via credit card and can be made monthly or annually.

2. Purchase through a G Suite Partner: Billings are invoiced, and payments can be via check or credit card.

For more specific information about payment plans including pricing differences for monthly vs. yearly options visit the website below:

https://gsuite.google.com/pricing.html

For businesses that want to purchase G Suite through a Partner, use the website below to find recommended Partners:

https://www.google.com/a/partnersearch/

Techniques for Email Transition

||||||||||||||||||||||||||||||||||||||||||||||||||||||||||||||||||||||||||||
Section Overview
There are times when a business is unable to migrate all employees to G Suite at once. How can your business gradually migrate email data from legacy systems to G Suite? How can you use a legacy system and G Suite in parallel?
||||||||||||||||||||||||||||||||||||||||||||||||||||||||||||||||||||||||||||

Three Deployment Phases

A typical G Suite deployment consists of three phases: Core IT, Early Adopters, and Global Go-Live. Each of these phases require a well planned messaging architecture to ensure a successful G Suite deployment.

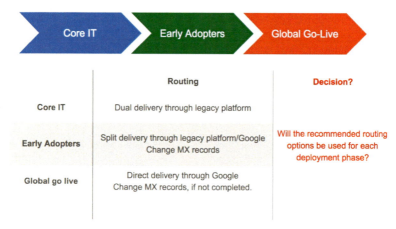

A typical G Suite deployment consists of three phases: Core IT Adoption, Early Adopters, and Go-Live. Each of these phases requires a well-planned messaging architecture to ensure a successful G Suite deployment.

Core IT

The legacy MX record will not change during core IT. Forwarding will be done via contact records or other appropriate measures. Users will send mail from the client in which they receive mail (only Core IT will send from Gmail)

Typically, many G Suite users refer to this topology as dual delivery. Depending on the legacy platform, they often have the choice of whether to deliver mail to both systems or not.

Shadow domains are always of concern for customers and should be addressed. There is no impact to existing mail-flow. MX records will always point to Google. Shadow domains are necessary because legacy mail systems will retain ownership for all addresses in a domain that it has been configured for. A shadow domain should be used for internal routing purposes only, not for email addressing purposes. Take care to prevent shadow addresses from entering the legacy system's Global Address List.

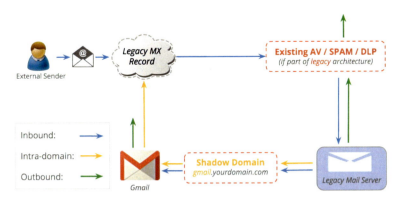

All inbound messages for your primary domain continue to be delivered to the user's mailbox on the legacy mail server. A copy of each message is forwarded to G Suite users via the shadow domain gmail.yourdomain.com.

Early Adopters

During Early Adopters, you should discuss inbound mail flow first, intradomain second, and outbound mail flow last. The legacy MX record will change prior to this phase. Users being added to Google will have their contact records (or other forwarding means) updated prior to MX change. Users will send mail from the client in which they receive mail (Core IT & Early Adopters will send from Gmail).

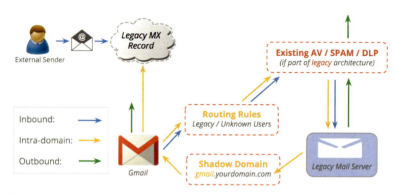

Mail routing is configured to route inbound messages from the external sender to either G Suite or the legacy mail servers. The MX records for the primary domain is changed to point to G Suite.

Global Go-Live

Early Adopters should be as much of a test run for Global Go-Live as possible. It makes the assumption all users will reside in Google at the time of Global Go-live. User should use the second animation as an "if needed" type of scenario for re-routing addresses that will not live in Google for any reason.

Outbound mail from Gmail can be directed through the existing

Data Loss Prevention (DLP) rules if it is necessary to do so. However, the existing DLP rules should be implemented directly into Gmail if possible.

All users are using G Suite. All distribution lists and resources have been moved to G Suite.

Data Migration from Legacy Systems

||||||||||||||||||||||||||||||||||||||||||||||||||||||||||||||||||||||||||||
Section Overview
Google offers free tools to migrate emails, contacts, and calendar data stored on legacy systems. Let's take a closer look at these tools and determine which is most applicable to your business.
||||||||||||||||||||||||||||||||||||||||||||||||||||||||||||||||||||||||||||

Data migration led by system administrators

To support data migration from major legacy systems, Google has developed migration tools that you can use for free. For example, to migrate data from Exchange or Lotus Notes/Domino, you can use G Suite Migration for Microsoft Exchange (GSMME) or G Suite Migration for IBM Notes (GSMIN), respectively.

Using these tools, you can easily migrate emails, contacts, and calendar data. If you are using a Microsoft Exchange Server, you can even migrate public folders.

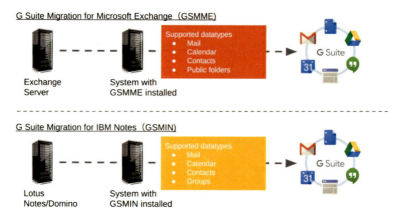

Both tools help transfer data such as emails, calendars, and contacts from legacy servers to G Suite.

GSMME also has IMAP migration functionality, which enables you to transfer mail data to G Suite from sources such as the consumer version of Gmail, rental servers, or ISPs.

GSMME enables you to migrate data from any IMAP capable server such as the consumer version of Gmail or rental servers.

Chapter 2: Getting Started with G Suite

If your company uses Microsoft Outlook and has information stored in PST files, you can use the GSMME tool to migrate your PST data to G Suite.

A PST file is a personal storage file in Microsoft Outlook. Using GSMME, you can migrate several PST files simultaneously to G Suite.

In addition to the tools listed above, within the G Suite Admin Console there is a "Data Migration Service", which enables you easily migrate mail and contact information. The service is entirely cloud based, so there is no need to install a migration client.

Using only a web browser and the G Suite Admin Console, system administrators can automatically connect to most email servers after entering basic information. The Data Migration Service can also migrate data from the consumer version of Gmail or another G Suite domain.

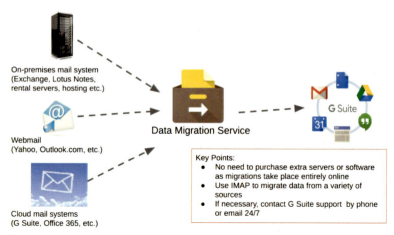

Within the G Suite Admin Console, there is a data migration service provided free of charge. The data migration service can pull data from various mail systems including on premises, web mail, and/or the cloud.

Data migration led by users

If you use Microsoft Outlook, it is possible to migrate data yourself as opposed to having it done by a system administrator.

Google has developed two tools to enable you to transfer your own data depending on your needs: G Suite Migration for Microsoft Outlook (GSMMO) and G Suite Sync for Microsoft Outlook (GSSMO).

If you won't be using Outlook after migrating your data, you should use GSMMO to transfer your data. On the other hand, if you want to use both Outlook and Gmail, you should utilize GSSMO to sync your data.

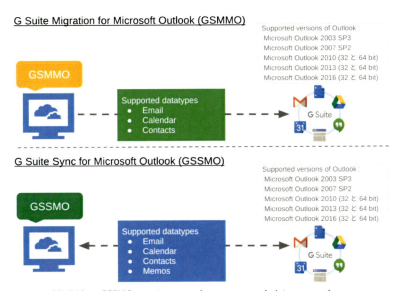

By using GSMMO or GSSMO, you (as opposed to a system admin) can transfer or sync your own data from Outlook.

Link your company's directory service to G Suite

To help companies migrate data to G Suite from a Microsoft Active Directory or LDAP server, Google provides a free tool called Google Cloud Directory Sync (GCDS). GCDS enables you to add, modify, delete, and synchronize data to help manage a large number of users.

GCDS performs a one-way synchronization from the directory server to G Suite. This means that you only need to manage user information in your LDAP server, and changes will be automatically synced to G Suite.

Also, if you would like to use the same passwords for both Active Directory and G Suite, you can use a tool called G Suite Password Sync (GSPS). Similar to GCDS, the synchronization is one-way. Therefore, whenever a user's password is changed in Active Directory, it will automatically be pushed to G Suite.

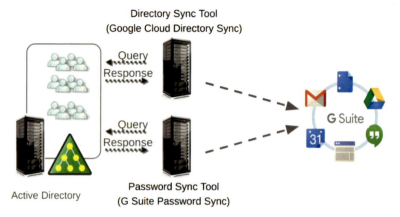

By using both Google Cloud Directory Sync (GCDS) and G Suite Password Sync (GSPS), your organization can use the same user IDs and passwords for Active Directory and G Suite.

Chapter 3: Change Management

Introducing the importance of Change Management

IIIIIIIIIIIIIIIIIIIIIIIIIIIIIIIIIIIIIIIIIIIIIIIIIIIIIIIIIIIIIIIIIIIIIIIIIIIIIIIIIIIIIIIIIIIIIIII

Section Overview

This section introduces the concepts of change management and explains why it is a critical part of every G Suite deployment project.

IIIIIIIIIIIIIIIIIIIIIIIIIIIIIIIIIIIIIIIIIIIIIIIIIIIIIIIIIIIIIIIIIIIIIIIIIIIIIIIIIIIIIIIIIIIIIIII

What is Change Management?

Change management is the workstream of a G Suite deployment that involves people and process related activities.

Change management will provide the support your employees need to successfully adopt the new technology and begin working in new, collaborative ways in order to help the organisation achieve its digital transformation objectives.

Deploying G Suite across your organization is a change project that will impact every employee. Typically people find change difficult, especially in the workplace, where for most people, the decision to change was not their own decision. Each individual will embark on their own personal change journey, and will likely experience a mix of emotions along the way. This is completely normal and to be expected. Your change management program will help people navigate this journey. See below for the typical stages, emotions that will likely be experienced, as well as a suggestion of the type of help that people need along the way.

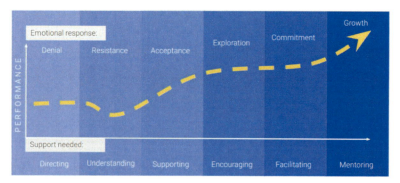

Stages of Personal Transition

The importance of Change Management

A successful G Suite deployment will result in your employees utilizing the full suite of tools (Forms, Sheets, Slides etc) to streamline their business processes, and improve efficiency, flexibility and collaboration. To get to this stage people will need to understand why using the tools will benefit them personally, and also how it will help the organisation to achieve its business objectives. People also need new skills, which will come from the training that is offered.

There are many additional benefits of managing change well - even beyond achieving project success. The risks of neglecting to manage change can be significant and will also impact more than just the success of this project.

Benefits of managing change	Risks of not managing change
• Project success • Productivity gains • Increased employee engagement • Improved collaboration and teamwork • Improved organisational communications	• Project delays • Employee resistance • Employee confusion / frustration • Lost productivity • Morale dip • Loss of trust of leadership among employees

Benefits of managing change properly VS the risks of not

According to 'The Value of Change Management' report conducted by Google and Raconteur in 2016, organisations who have an excellent and sophisticated change management program in place achieve +30% ROI than those without. In the next section we'll discuss the elements that make up Google's proven change management methodology.

Google's proven Change Management Methodology

||||||||||||||||||||||||||||||||||||||||||||||||||||||||||||||||||||||||||||||||||||||||||||||
Section Overview
This section provides an overview of Google's proven Change Management Methodology.
||||||||||||||||||||||||||||||||||||||||||||||||||||||||||||||||||||||||||||||||||||||||||||||

Overview of the methodology

The Google Change Management Methodology was designed to support employees as they transition from their existing IT tools over to G Suite. The focus of the approach is to ensure that employees understand why the company has decided to go Google, what the benefits are for the organisation as well as for them as individuals and to give people the skills and ability to be successful. There are four workstreams that make up the methodology:

4 critical elements in Google Change Management Methodology

1. Sponsorship & Engagement:

The single most influential factor to the change project being successful is having strong, active, visible executive sponsorship from the CEO and other business leaders. This is so that employees understand and commit to this being a critical business project with desirable business outcomes, rather than simply 'something that the IT team is working on' which will typically have less ability to engage employees. The executive sponsor has a critical role in communicating to all employees about why the change is happening and how it aligns with the organisations business goals. The earlier you engage business leaders in the project the better. They'll have useful input into how the project should be executed, and this participation in the early stages will help ensure they feel a sense of ownership and are committed to project success.

Another important activity that takes place during the 'Sponsorship and Engagement workstream' is the creation of a champions network. You should recruit 5-10% of your employees to become champions. Ideally these should be volunteers who represent each of your major teams, divisions and locations. They will go-live on the platform in advance of the rest of employees. This will allow you to test and iterate your communications and training plans, and to get champions upskilled, so that they can provide peer to peer support to the rest of your employees during 'global go-live'. The role of champions should be clearly defined and will include activities such as sharing project updates with their peers, providing the project team insights into the employee sentiment in each team and location, and providing coaching and help to their peers during go-live. It is important that the champions are well respected members of their teams and have good communication and interpersonal skills so that they can inspire and help others and represent the project in the right way. It is also both useful and fun to create a brand for your champions and make them visible to the rest of the organisation. This could be as simple as giving them an interesting name (e.g. the Google Hero's or Gurus) and getting them to wear a recognisable t-shirt on go-live day. Creating a community (Google+ or via Google Groups) will allow champions to share their best practices, tips and success stories with others in different regions and teams. You might also want to request a small budget that can be used to reward high performing change champions. The above recommendations will help you to fully engage your key stakeholders throughout the deployment helping to smooth the execution of the change management plan.

2. Organizational Analysis:

This is the body of work that needs to happen in order that you clearly understand the impact and benefits of the change on the different employee groups within the organisation. Having a clear understanding of this will allow you to develop compelling and relevant communications and training plans. To do this you will need to collect data on the following:

What are the different types of employee groups that exist within the organisation?

How many people are there in each group?

Which locations are they based and which languages are they using?

How does each user group currently use their existing IT tools in their day-to-day business operations and how will this change once they move to G Suite?

Which critical business processes have heavy dependency on existing IT tools and will be impacted by the move to G Suite?

What communications channels exist to connect with each group?

What is the best format for people within each group to be trained and develop the new skills required?

Your champions will be able to help capture much of the above information about different employee groups, but you will also need to work closely with your HR and IT teams. You might also consider sending surveys or running focus groups with different teams to capture the required level of detail needed. This information will allow you to tailor communications to the specific needs and language requirements of employees, and will also give you the information needed to design a training program that will be high impact and minimise the chance of people being training on topics not relevant for them - which can be wasteful of training budget or resources.

Another activity to complete within the Organisational Analysis workstream, is to consider how you will measure the impact and success of your project. Organisations typically look at the metrics including usage level (how many '30 day active users' there are for each of the products), the impact of process improvement (amount of time or money that has been saved), the level of employee engagement in the project (attendance rates of training sessions, number of employees are visiting the project website, number of people providing feedback to the project team) and employee satisfaction (ease of use of tools, improvements in collaboration and flexible, satisfaction with training). Once you've completed the above work, you'll be able to start work on both the communications and training workstreams.

3. Communications:

The project needs a clear vision statement or 'elevator pitch' so that every employee clearly understands why the project is happening and what the benefits of this change are. The elevator pitch should be a short and simple, but meaningful statement that is shared during all communications and training activities. There should be no excuse for your employees not knowing 'why' they are expected to change. An example of an elevator pitch for a made up company is below:

Elevator pitch example

Elevator pitch for ZoomZoomMedia

We are very excited to announce that ZoomZoomMedia has made the decision to switch to G Suite for our day to day business tools. This will mean each and every one of you is going to have access to fantastic Google products that will allow us to work more collaboratively, to get some of our boring tasks done much more quickly and to truly enable flexible working for everyone. The time we save means that we can give even more focus to our customers and go big on our innovation dreams. We can't do this without you - so I hope you're onboard! Lots of training and fun activities are planned - we can't wait to get started on the next phase of ZoomZoomMedia's exciting journey.

Mary CEO

A typical mistake that organizations make is under communicating to their employees. When you are at the point of thinking you've communicated too much - you're probably only just starting to get through to people. There is no such thing as over communicating ! Do not rely on email as your only communications channel. Consider using team meetings, company blogs, posters, newsletters, and any other channels that might exist. Communications materials should be designed and branded in a fun and engaging way so that people pay attention to them. Ensure that email communications are sent by someone senior and well known within the organisation (preferably the executive sponsor) to ensure that they are read. You might find ways to allow employees to participate in the design and development of the project branding - this will increase their engagement and involvement of the project, and also mean you're likely to get some creative and interesting ideas.

4. Training:

One of the main things that people fear about change is fear of looking incompetent. Employees must be trained on how to use the new technology in order that they develop the new skills they need to feel confident, happy and efficient. There are a variety of training formats that you might consider offering to your employees. Given that your employees will have a mix of learning styles the best idea is to offer multiple training formats. Suggested formats include self-paced learning (videos, handouts and e-learning) and instructor-led training (remote via Google Hangouts, or in-person). Depending on the learning objectives and locations of each of your employee groups you might end up offering all employees self-paced options and then a selection of important groups receive instructor-led training. Planning and delivering training can take time, so it is best to start early to ensure it has maximum impact. Training should be offered both before and after the go-live date as people will likely have different questions at each stage.

Top tips for any instructor-led training include:

- Sharing the elevator pitch to remind people of the importance of the project and why training will help them be successful
- Focus on the identified use cases that are most relevant for the group. It will be a waste of time to train all employees on all tools and all features. Rather focus on the things most important and beneficial for them specifically.
- Where possible deliver hands on training so participants get the opportunity to practice in the classroom. This will increase the amount of new information that they are able to retain once they

step back into their day.

Additional activities

Additional activities that you should consider include celebrating go-live with a party or event. This will bring awareness to the project and encourage people to get involved. The celebration can be as simple or elaborate as you like, the important thing is that it is recognised as the beginning of a new way of working. It is also important to reward and recognise people that embrace the new ways of working. This will help inspire others to get involved. Again, this does not need to cost money - and can be as simple as public recognition from leadership or a certificate awarded to the deserving employees.

Our final suggestion for this section is 'don't let up too soon!'. One of the most common reasons that change projects are not as successful as they could and should be is that project efforts are stopped too early. It is important to maintain energy around the project beyond the go-live day. Our suggestions for how to do this are captured in the next section.

Increasing adoption of G Suite post go-live

||||||||||||||||||||||||||||||||||||||||||||||||||||||||||||||||||||||||||||||||||||||||||||||||||||

Section Overview

This section provides tips and best practices to increase adoption and the positive impact of G Suite post go-live.

||||||||||||||||||||||||||||||||||||||||||||||||||||||||||||||||||||||||||||||||||||||||||||||||||||

The go-live date is just the beginning of the journey

Every organisation has different reasons for wanting to deploy

G Suite. Many aim to improve how employees communicate and collaborate, some focus on wanting to increase productivity or efficiency, and others want to provide employees with a platform that will help inspire a more innovative culture within the organisation. Go-live day is just the beginning of an organisation's journey to achieve transformation. Ongoing support for employees is critical to embed new behaviours and new ways of working. Below we make suggestions for how to continue your transformation journey.

Ongoing organisational analysis

To make continual improvements in the way that employees use the tools and to increase the value they bring to your organisation it is important to regularly check in to understand the current status of adoption and employee sentiment. Collecting survey data can be a useful way of doing this and will highlight any areas that might need some new or additional communications or training activities.

Facilitate Transformation Labs

Your post go-live plan should include the facilitation of 'Transformation Labs'. A 'Transformation Lab' is typically a half or full day workshop that involves a four stage journey: Inspire, Explore, Prototype and Roadmap. The approach has been developed using design thinking principles, and allows employees to solve their own business problems through a series of interactive exercises. Labs will help employees uncover new and exciting ways to use the technology in order to improve efficiency, collaboration and to increase innovation. These include simple use cases such as compiling data or information on a single document or slightly complex use cases such as automating business processes as well as integrating existing applications with G Suite by using Google Apps Script.

Transformation Labs

Establish an innovation council

To ensure that change is fully embedded and a culture of innovation is encouraged to thrive, we suggest that you establish an innovation council. This council will be made up of a mix of your champions as well as other leaders and decision makers. Their responsibilities include sharing best practices and success stories between regions and teams, and identifying new opportunities to improve business processes. The council can also help identify any specific training needs that might exist within their specific team or region.

Continue with communications and training activities

Employees will not become product experts overnight, so it is important to continue to offer training after go-live. Additionally you must also consider how to train new employees that join the organisation who maybe haven't worked with G Suite previously. You'll also find that there are regular product enhancements and new features that are made available - so giving employees training on these updates will help them to get further value from the technology and to improve additional processes.

Use your communications channels to share success stories and use cases that have been identified. If they've worked in one part of your business, maybe you can share more widely and get even more value.

Identify where you are on your Transformation Journey

Based on our experience of having worked with thousands of organisations across the world on their digital transformation journeys we've identified a number of stages that organisations typically move through. Details of these stages are shown below.

G Suite Organization Transformation Journey (based on Gartner's Maturity Model for Enterprise Collaboration and Social Software)

Chapter 3: Change Management | 83

The curve across the top half of the above diagram shows how individuals within an organization typically use G Suite as they become more advanced. The stages in the bottom half of the diagram show the impact that an organization should see as a result of the increase in adoption and changes in employee behaviour. When a minimal number of employees are using G Suite with no or only basic level of collaboration, an organization will only see benefits among limited group of people. For example, individuals might appreciate being able to access work documents from anywhere with any device, allowing them more flexible working hours. Or small groups of people might collaborate on a document and increase their efficiency. However when there is widespread G Suite usage among employees, the organization will see an increase of benefits at team, department and organization wide level. For example having multiple groups proactively improving their existing business processes by using G Suite. Or company wide communications becomes more interactive and impactful. Identify where your organisation is on this journey, and strive to move to the next stage through the activities we've mentioned above.

Project execution

In order to ensure your project is successful you'll need to assign a change management lead who is responsible for driving the activities detailed above. This person will need to work very closely with other members of the deployment team, both in the project management (governance) workstream and on the technical workstream. This is to ensure that dates are aligned and that employee data is migrated across per the deployment blueprint that has been defined during your deployment planning workshop.

Attributes of a good change manager

The change manager will be responsible for creating and delivering the change management plan before and post go-live. The ideal candidate for such a role will have the following attributes:

a. Experience and knowledge of change management principles, methodologies and tools

b. High levels of empathy and emotional intelligence

c. Passion energy and enthusiasm for new things

d. Exceptional communication skills – both written and verbal

e. Ability to establish and maintain strong relationships with people across all levels of an organisation

f. Flexible and adaptable; able to work in ambiguous situations

g. Organised and able to deal with and create complex project plans for large projects

h. Problem solving and root cause identification skills

i. Team player and able to work collaboratively with and through others

j. Business acumen and understanding of organisational issues and challenges

k. Familiarity with project management approaches, tools and phases of the project life cycle

If you don't have such a person within your organisation, you might consider hiring someone, or working with a Certified Google Partner. We have a network of partner organisations that have been through detailed training on how to deliver the above methodology and who have experience and expertise in helping other organisations embark on this journey.

Chapter 3: Change Management | 85

Chapter conclusion

|||||||||||||||||||||||||||||||||||||||||||||||||||||||||||||||||||||||||||||||||||||||

Section Overview

This section is the conclusion for this chapter.

|||||||||||||||||||||||||||||||||||||||||||||||||||||||||||||||||||||||||||||||||||||||

Closing note

Change Management is really critical if you want your project to be successful. It will inspire and motivate your employees and ensure that they are set up for success in the new world. It will help your organisation work more efficiently, increase collaboration and build agility into the organisation for the future. The only constant in life is change - so you may as well do it well!

Big thanks to the colleagues that helped us write this chapter, specifically Katia Stapley-Oh for her contribution of the Transformation Journey Model.

Chapter 4: Use cases from existing customers

How companies are leveraging G suite to transform productivity

||||||||||||||||||||||||||||||||||||||||||||||||||||||||||||||||||||||||||||||||||||||||||||||||||||||||||||

section overview

In this section we will be sharing and discussing how companies across the world are leveraging g suite to drive greater productivity and efficiencies.

||||||||||||||||||||||||||||||||||||||||||||||||||||||||||||||||||||||||||||||||||||||||||||||||||||||||||||

Security enhancement

One of the largest ecommerce in asia started off with G Suite when the company was first incorporated. They were also a digital native company and really loved G Suite's "Serverless / No Ops" concept, where they no longer had to procure, build and maintain file and email servers. G Suite was deployed as the main tool for email communication.

The company quickly grew from a few dozen people into a nearly 2,000 people company, becoming the largest B2C eCommerce company in the region. Due to the nature of their business they had to handle large number of customer's data. These were very sensitive data such as credit card information, that were tied to mobile phones and personal identification details such as names, and date of birth. This could have very serious repercussion if the information were leaked outside of the organisation.

In order to safeguard the data privacy of their customers against intentional or unintentional leakage of data, the customer deployed G Suite Data Loss Prevention (DLP) across their email and drive communication. They love the Optical Character Recognition capabilities of the G Suite platform, which incorporated Machine Learning capabilities into the DLP function. With G Suite, the company has the means and capabilities to ensure a strong information privacy control, and to meet the highest security standards required by regulators.

Legacy migration

A very large publishing company in Asia wanted to explore ways to reduce their cost of IT operation from their current Legacy environment. The management of the company wanted to look for possible avenues in terms of cost savings and driving greater productivity through Business, Cultural and IT Transformation.

They found out about G Suite and realised quickly the enormous benefits. First, G Suite offered them to move from an "Asset-heavy" IT infrastructure world into an "Asset-light" Infrastructure world. With G Suite, they no longer had to purchase any more servers, recruit more people to manage their email, video conferencing, productivity and collaboration environment. They also wanted to leverage the G Suite platform to enhance and empower their "Field workers" - the reporters and correspondent across the world - to collaborate and share information in a safe and secure manner.

They were also amazed at G Suite's App Maker (At the time

of writing, this is in Early Adopter Program). They saw how quickly they were able to build and deploy simple applications internally within their organisation by empower every functions and divisions to develop applications through App Maker.

Global Communication Platform

A global leader in research-focused healthcare with combined strengths in pharmaceuticals and diagnostics moved to G Suite.

For the last two and a half years, they maintained two different email and calendaring platforms, which had often been an obstacle for effective collaboration. To end these platform interoperability issues, the IT Steering Committee made the decision that all employees will move to G Suite as the single common platform for their entire Group.

When they evaluated new cloud-based solutions, the IT Steering Committee was impressed with the outstanding service and rapid innovation of G Suite. G suite will enable over 90,000 employees of their total employees base to work better together from anywhere.

The way their employees communicate and collaborate is diverse, and their employees are spread across over 140 countries. The integrated and socially-focused way that G Suite enables collaboration is very compelling for them, and they expect this to not only bring their company closer together, but to give them a strategic advantage. Additionally, being able to deploy G Suite by simply enabling them via a control panel versus planning for and deploying complex infrastructure in their datacenters will help them focus on their core business -- helping save patients' lives

Employees will be able to access their email and documents from any web-enabled device, without using remote access systems such as VPN. This will make it easier for employees to work from home or on the go and it will reduce the strain on IT support teams. Removing barriers to communication and innovation while enhancing mobile access is a key part of their IT strategy.

Sales Forecast with Google Sheets

A large manufacturing company in India has been using G Suite for the last few years and they use most features of G Suite across the organization.

One key use case for them has been the process of using Google Sheets for annual sales projection and sales forecast. This specific company has many distributors across India.

Before they went on G Suite, one of the ops admin used to call each distributor every week and get the sales forecast. This was a highly manual and tedious process which used to take a few hours. Furthermore, as it was a manual process it was prone to errors.

The customer calls G Suite a "game changer" for them.

Now they have a Google Sheets which they share with the distributors and they update them on real time basis. This has led to a much more efficient process for the customer wherein they have simplified annual sales projection and sales forecast by updating and tracking them on Google Sheets. This eliminated sending attachments back and forth and making updates on phone calls. Furthermore, the marketing team used Google Drive to make updating and sharing marketing collaterals and product catalogs across different locations effortless.

90 | Chapter 4: Use cases from existing customers

G Suite for Recruitment

A large global technology company uses G Suite for it's entire recruitment process.

First they use Google docs to create a job description. The recruitment team and the hiring manager collaborate together on a Google Docs to finalize the job description rather than sending attachments back and forth to finalize the job description. Then the recruiter creates a Google Sheets which is shared with the hiring manager and all details about the shortlisted candidates are mentioned in the Google Sheets including a link to the resume which is stored in Google Drive. Then the recruitment team starts scheduling interviews with the interview panel by checking their availability using Google Calendar. Finally, the interviews for remote candidates are conducted using Google Hangouts and the interviewer(s) update the relevant Google Sheets with their comments/recommendation for the candidate.

The usage of many features from the G Suite portfolio help this company to reduce the JD to Hiring cycle significantly thus driving more efficiency and increase in productivity.

Connecting with the organization socially

One of the largest supermarket chain in Australia uses G Suite to connect it's 100k+ staff with their peers and management team.

This company actively uses Google+ to update employees about the latest happenings at their various stores across the country. Employees also upload pictures and anecdotes from various events which are conducted at the various stores and then others chime in to provide comments and their suggestions. This drives a deeper sense of belonging for all the employees and helps them to connect with each other across different locations without the need for any additional investment. Store managers also jump on a hangout video call from any device on the fly to share best practices and collaborate on innovative ideas.

This has helped them to create a very well connected collaborative org where people can diffuse boundaries and hierarchies to connect and collaborate together seamlessly.

Chapter 5: Merits of Deploying G Suite

Business transformation with G Suite

||||||||||||||||||||||||||||||||||||||||||||||||||||||||||||||||||||||||||||||||||||||||||||||||

Section Overview

G Suite can transform the way your business works. More than three million businesses have adopted G Suite, and Google's own employees use it every day to drive innovation at their workplace. How can your workplace be transformed, and how does this transformation helps organizations' business success?

||||||||||||||||||||||||||||||||||||||||||||||||||||||||||||||||||||||||||||||||||||||||||||||||

Workstyle transformation with G Suite

In today's workplace, employees typically work individually at their computer, create a project or proposal, upload the file in an email or to a server, and then contact relevant team members.

G Suite makes this process remarkably more efficient. The foundation of a G Suite business transformation is information sharing. Sharing, itself, is the cornerstone of the change; it fosters a new (and refreshing) mentality of working together as opposed to working individually.

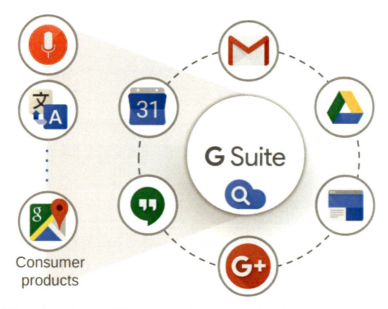

With only a web browser, G Suite provides various communication tools to enhance your workplace.

Efficient information sharing helps to achieve your business goals

How can workplace transformation help you reach your strategic goals? Every organization has its own goals, so let's look at a common goal for most organizations - increasing revenue. The answer is efficiency. For example, imagine a scenario where one of your sales representatives can't meet one of his/her customers. It would be great if another sales representative could fill in as a backup, but unfortunately, no one else is knowledgeable about the situation or the customer, and the meeting is canceled.

This is why information sharing is critical. If previous meeting notes, customer concerns, or contract details had been shared with other colleagues, another colleague may have been able to fill in to avoid a customer delay.

At the end of the day, there are many ways to increase revenue. The challenge is to do it efficiently, and G Suite can surely help your organization to achieve this without adding more resources to your business (See below for the example of how G Suite can help your organization to achieve business goals).

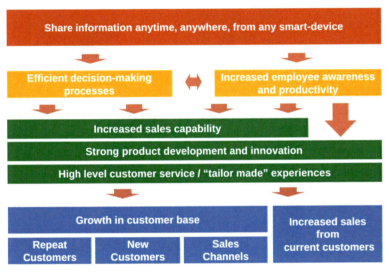

By enhancing your organization's communications capabilities, G Suite will ultimately help you reach your strategic goals.

Closing thoughts

G Suite can be described as the ultimate toolset for information sharing and effective communication. Whether your strategic goals are to increase revenue, decrease costs, or anything in between, your company will benefit from applying these services. Challenge yourself - see how G Suite can transform your business. The cloud is waiting for you, and it's only a click away!

About the Author

Yoshiki Sato

Yoshiki Sato is an experienced technology advisor and author. He worked at Hewlett-Packard, Microsoft and Google for over 16 years as a Solutions Engineer, Developer, IT consultant, and Product Manager, specializing in enterprise business and he is working for Google as a Technical Specialist Lead for Google Cloud.

Kim Wylie

Kim leads the Customer Change and Culture team within Google Cloud. She is an expert in change management, the psychology of change, digital transformation, organisational culture, leadership and high performing teams. She designed and developed the Google Cloud change and transformation methodology which has been used by tens of thousands of Google's customers globally. Kim is a regular keynote speaker on the topics of organisational change, innovation, and leadership and has been published in The Economist and Forbes. She is also a regular guest lecturer at London Business School.

Ayaka Yamada

Ayaka Yamada is an expert in organizational change management and digital transformation with enterprise cloud technology and has extensive experience of providing trainings, workshops on change management and digital transformation to numerous enterprise customers.

Kenneth Siow

Ken is currently the Head of Strategic Accounts Management for Google Cloud across the Asia markets, responsible for helping Customers to overcome their business challenges through the power of the Cloud and managing the current book of business in Google Cloud G suite existing base. Prior to Google Ken held a number of leadership roles including heading Cisco System's APAC Professional Services Partner Team, Services Consulting Sales for Intel Greater China, and Information Management Consulting lead at Royal Dutch/ Shell covering APAC and EMEA Markets, focusing on complex Management Information Systems/ Business Intelligence project implementation.

Shahid Nizami

Shahid leads the Account Management team for APAC at Google Cloud. He has more than 15 years of experience in various customer facing roles in the IT industry where he has been working closely with CIOs & CTOs across the APAC region with a very deep understanding of the functional relevance of technology in driving digital transformation. He has an MBA in Marketing & Systems and is very passionate about how technology is making the world a better place.